Dariusz Karnas

POLISH AIRCRAFT (1939) INSTRUMENT PANELS

Published in Poland in 2017
by STRATUS sp.j.
Po. Box 123,
27-600 Sandomierz 1, Poland
e-mail: office@mmpbooks.biz
for
MMPBooks,
e-mail: rogerw@mmpbooks.biz

http://www.mmpbooks.biz

ISBN
978-83-65281-40-1

Editor in chief
Roger Wallsgrove

Editorial Team
Bartłomiej Belcarz
Robert Pęczkowski
Artur Juszczak

Drawings
Dariusz Karnas

Proofreading
Matthew Willis

DTP
Stratus sp.j.

Printed by
Drukarnia Diecezjalna,
ul. Żeromskiego 4,
27-600 Sandomierz
www.wds.pl
marketing@wds.pl

PRINTED IN POLAND

Introduction

Since the beginning of manned flight, it has been recognized that supplying the pilot with information about the aircraft and its operation could be useful and lead to safer flight. Original analogue flight instruments are operated by air pressure and the use of gyroscopes.

There are three basic kinds of instruments classified by the job they perform: flight instruments, engine instruments, and navigation instruments.

Flight Instruments – the instruments used in controlling the aircraft's flight attitude are known as the flight instruments. There are basic flight instruments, such as the altimeter that displays aircraft altitude; the airspeed indicator; and the magnetic direction indicator, a form of compass. Additionally, an artificial horizon, turn coordinator, and vertical speed indicator are flight instruments present in most aircraft.

The centre position directly in front of the pilot and co-pilot is the basic display position for the integrated flight controller (airspeed indicator, turn and bank indicator) or artificial horizon.

Engine instruments are those designed to measure operating parameters of the aircraft's engine(s). These are usually quantity, pressure, and temperature indications. They also include measuring engine speed(s). The most common engine instruments are the fuel and oil quantity and pressure gauges, tachometers, and temperature gauges.

Navigation instruments are those that contribute information used by the pilot to guide the aircraft along a definite course.

The instruments' badges in Polish Military Aviation had different colours.

Flight Instrument's (airspeed indicators, variometers) badges were dark blue.

Engine instruments:

- Oil pressure and temperature gauges were brown,
- Fuel pressure and fuel contents gauges was dark red;
- Boost indicators were red
- Tachometers were black.

Acknowledgements

The author wishes to acknowledge the kind assistance of the following: Bartłomiej Belcarz, Andrzej Glass, Tomasz Kopański, Marek Rogusz, Katarzyna Zielińska from Polish Aviation Museum-Kraków.

PZL P.11c

The PZL P.11 was a Polish fighter aircraft, designed in the early 1930s by PZL in Warsaw. It was briefly the most advanced fighter aircraft of its kind in the world. The PZL P.11 served as Poland's primary fighter defence in the Polish campaign of 1939, but with the rapid advances in aircraft design in the late 1930s (seen in such fighters as the Messerschmitt Bf 109), it proved outclassed by its rivals at the onset of the war.

The aircraft was conventional in layout, with high wings, all-metal, metal-covered. The cockpit was open. An internal fuel tank in the fuselage could be jettisoned in case of fire or other emergency. The armament was two 7.92 mm machine guns on fuselage sides, though a third of the P.11cs had two additional machine guns in the wings. The P.11c could carry four small 12.5 kg bombs (P.11a could not). A radial engine was used: Bristol Mercury V S2 (565 hp, max: 600 hp) or Mercury VI S2 (590 hp, max: 630 hp).

Right: *PZL P.11c cockpit, photo from Technical Manual.*

Below: *PZL P.11c serial no. 8.14, code no. 1 of* kpt. *Stanisław Pietraszkiewicz, 112 EM (*Eskadra Myśliwska – *Fighter Squadron), May 1936. (via T. Kopański)*

PZL P.11c

CIŚNIENIE
ŁADOWANIA
G.GERLACH
BENZYNA
SMAR
TEMPERATURA
PALIWO
CIŚNIENIE
SMAR
TEMPERATURA
SMAR
CIŚNIENIE
BENZYNA
SZYBKOŚĆ
SKRĘT
OBROTY
WYSOKOŚĆ
×100 m

1. Master compass
2. Integrated flight controller (Airspeed indicator, Turn and bank indicator)
3. Boost pressure gauge
4. Output oil temperature gauge
5. Fuel pressure gauge
6. Tachometer
7. Fuel contents gauge
8. Ignition switch for magnetos panel
9. Input oil temperature gauge
10. Oil pressure gauge
11. Fore-and-aft level indicator
12. Altimeter
13. Board clock
14. Emergency lever
15. Fuel tank selector
16. Fuel tank lever

Output oil temperature gauge

Above: *Boost pressure gauge*

Left: *Integrated flight controller (Airspeed indicator, Turn and bank indicator)*

Below: *Tachometer*

Fuel pressure gauge

Fuel contents gauge

Input oil temperature gauge

Oil pressure gauge

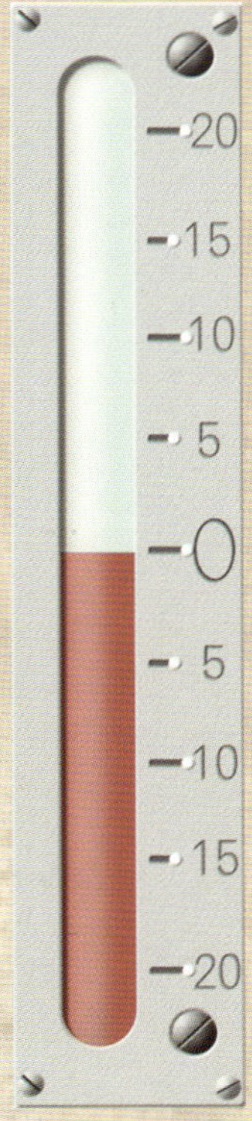

Fore-and-aft level indicator

Altimeter

Board clock

Above: *PZL P.11c instrument panel and windshield.*

Right: *PZL P.11c starboard side of the canopy, control column is visible.*

Below: *Starboard side of the fuselage with panels removed, machine gun is visible.*

(All photos from Technical Manual)

PZL.37 Łoś

The PZL.37 Łoś (in Polish – Moose) was a Polish twin-engine medium bomber, used in the defence of Poland against the Nazi German Invasion in 1939.

The Polish Air Force started to receive the Łoś A in early 1938, with deliveries of the PZL.37B to operational units, slowed by delays in supply of the Pegasus XX engines and other equipment, in the autumn of 1938. On 1 September 1939, it had about 86 PZL.37s in total, but less than a half of those were used in combat. 36 PZL.37Bs were in four bomber *escadras* of a Bomber Brigade: the 11th, 12th, 16th and 17th *escadras* (two *escadras* with nine aircraft each, constituted a group, in Polish: dywizjon; the PZL.37 were in groups X and XV).

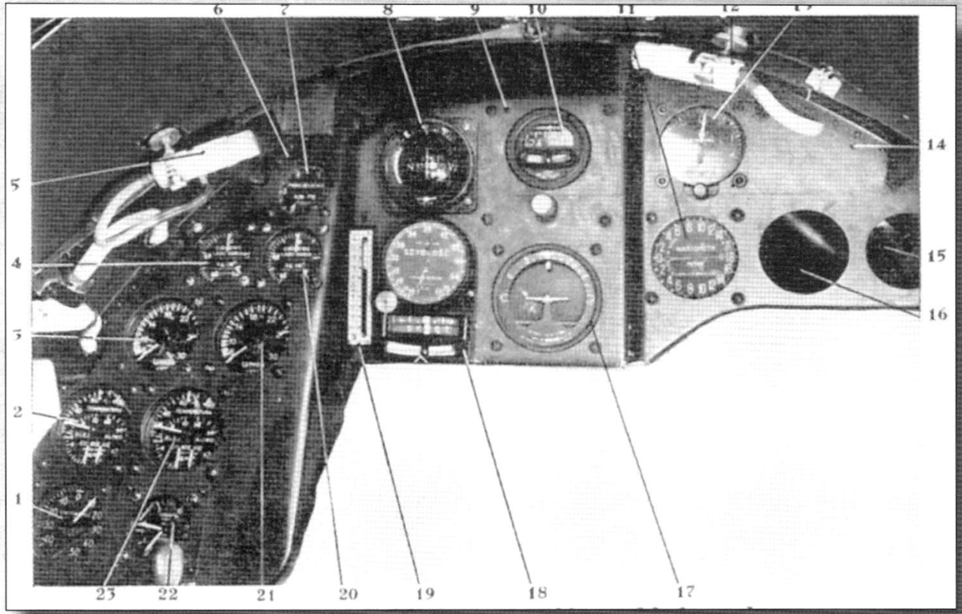

Right: *PZL.37 Łoś instrument panel, photo from Technical Manual.*

Below: *PZL.37 Łoś A bis, serial production. (A Glass coll.)*

PZL.37 Łoś

Watch interactive version of the panel.
http://mmpbooks.biz/Los_panel/PZL%2037-instrument-panel.html

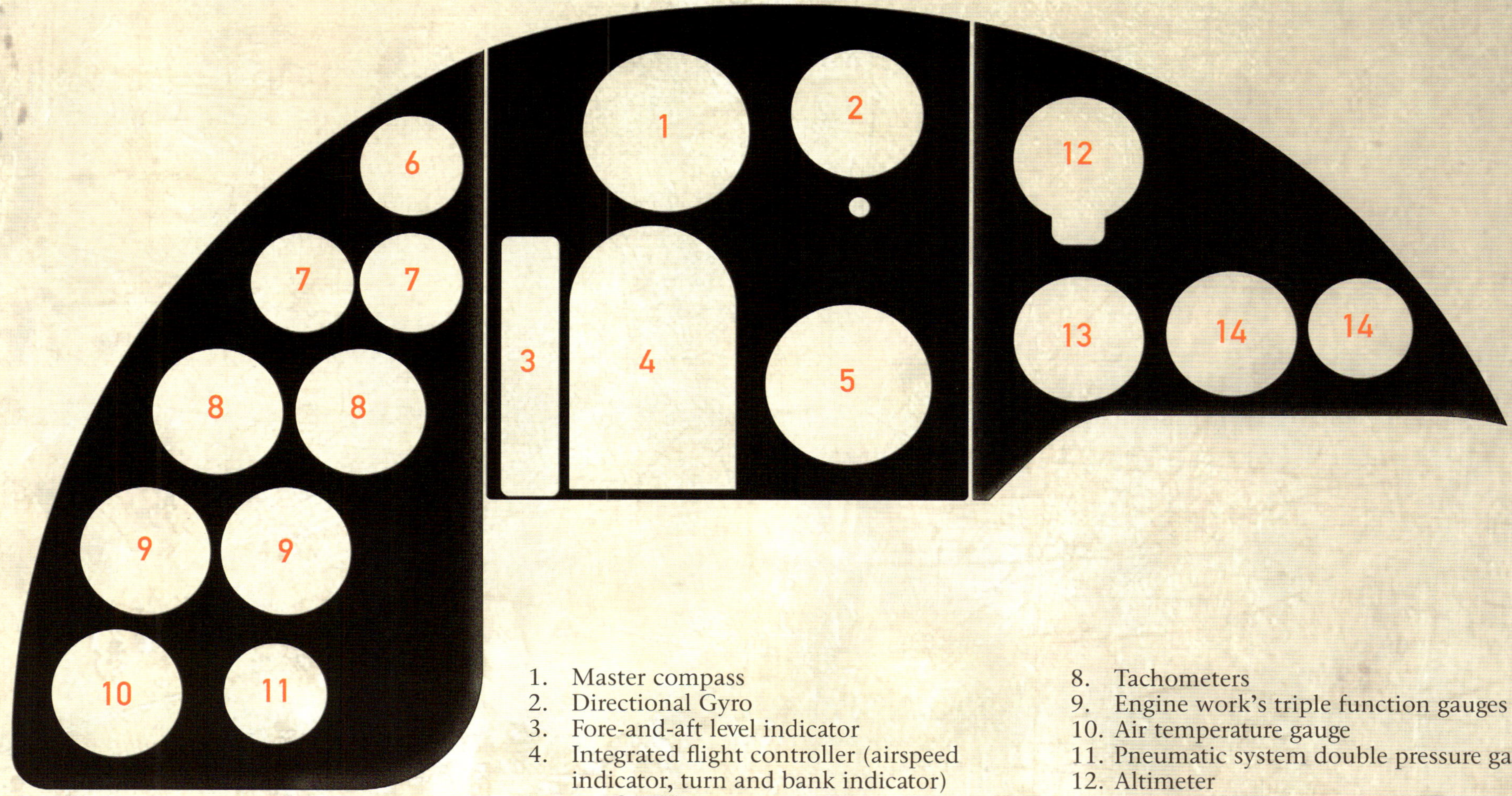

1. Master compass
2. Directional Gyro
3. Fore-and-aft level indicator
4. Integrated flight controller (airspeed indicator, turn and bank indicator)
5. Artificial horizon
6. Vacuum installation pressure gauge.
7. Boost pressure gauges
8. Tachometers
9. Engine work's triple function gauges
10. Air temperature gauge
11. Pneumatic system double pressure gauge
12. Altimeter
13. Variometer
14. Auxiliary slots for other instruments

Above: *Master compass*

Left: *Fore-and-aft level indicator*

Above: *Artificial horizon*

Right: *Integrated flight controller (airspeed indicator, turn and bank indicator)*

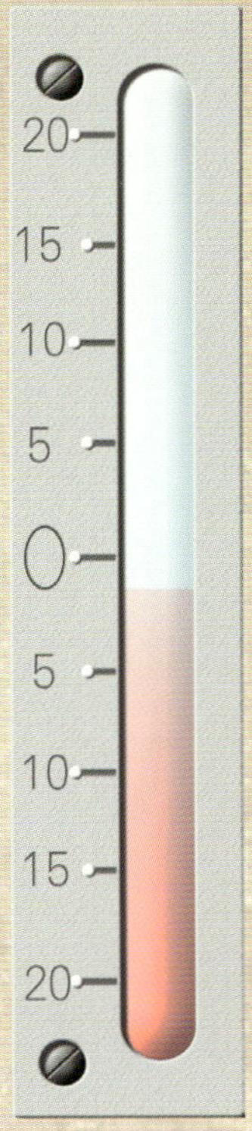

Vacuum installation pressure gauge

Boost pressure gauges

Directional gyro

Tachometer

Above: *Engine's triple function gauge (Temperature, oil and fuel pressure)*
Below: *Altimeter*

Above: *Outside air temperature gauge*
Below: *Variometer*

Pneumatic system double pressure gauge

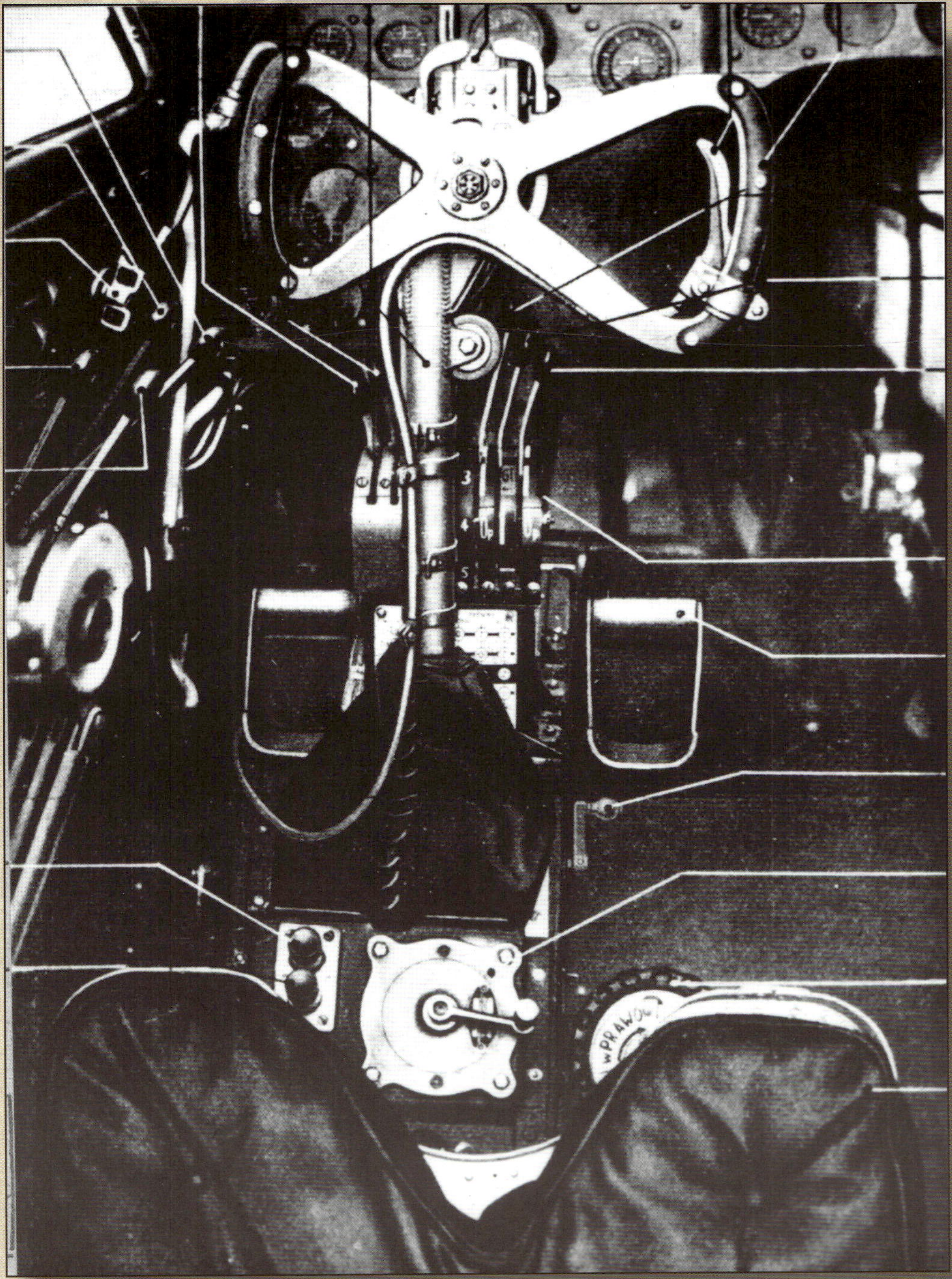

Left: *PZL.37 Łoś control column, photo from Technical Manual.*

Below: *PZL.37 Łoś front gunner position. (T Kopański coll.)*

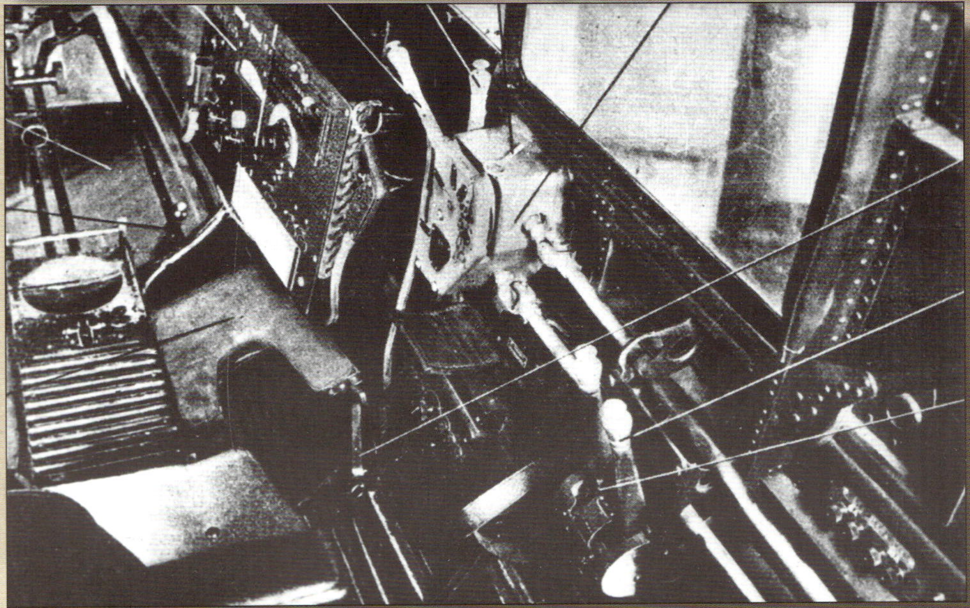

Bottom: *PZL.37 Łoś starboard side of the main cockpit, photo from Technical Manual.*

PZL.23 Karaś

The PZL.23 Karaś (in Polish – the crucian carp) was a Polish light bomber and reconnaissance aircraft designed in the mid-1930s by PZL in Warsaw. It was the primary Polish reconnaissance bomber in use during the Invasion of Poland.

The aircraft was a low-wing cantilever monoplane of all-metal, metal-covered construction. The crew of three consisted of a pilot, a bombardier, and a rear gunner. The bombardier's combat station was situated in a gondola underneath the hull, where he could also operate an underbelly machine gun. The fixed undercarriage was well spatted, but despite a massive look, it was not suited for rough airfields. Bombs were carried under the wings: the maximum load was 700 kg (1,500 lb) (6 x 100 kg and 2 x 50 kg).

***Right:** PZL.23 Karaś front instrument panel, photo from Technical Manual.*

***Below:** PZL.23 Karaś serial no. 44.62, code no. 9 of the 22 EB (*Eskadra Bombowa – *Bomber Squadron), 2* Pułk Lotniczy *(Air Regiment) in Cracow. (T. Kopański coll.)*

PZL.23 Karaś

Klapy otwierać przy predkości miniejszej 160 km godz
SZYBKOŚĆ
SKRĘT
WYSOKOŚĆ
km
×100 m
w górę
w dół
m/s
SPERRY
DIRECTIONAL GYRO
G. GERLACH WARSZAWA
POWIETRZE
CIŚNIENIE
SMAR
CIŚNIENIE
CIŚNIENIE ŁADOWANIA
OBROTY
OLEJ
TEMPERATURA
PALIWO
CIŚNIENIE
POCIĄGNĄĆ POKRĘTŁO ZABLOKOWANE OLEJ
BENZYNA
POCIĄGNĄĆ POKRĘTŁO ZABLOKOWANE CHŁODNICA

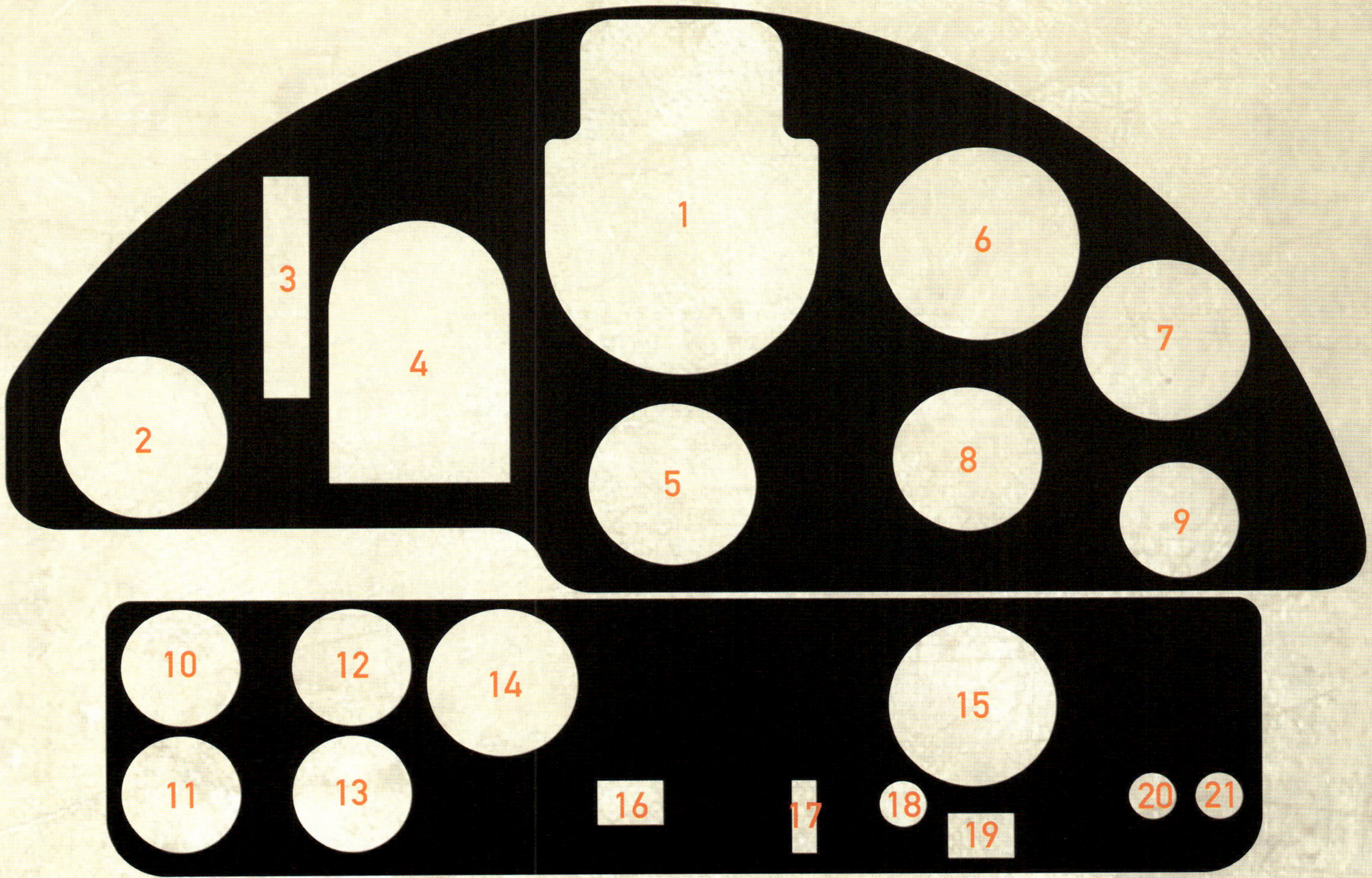

1. Master compass
2. Altimeter
3. Fore-and-aft level indicator
4. Integrated flight controller (airspeed indicator, turn and bank indicator)
5. Precision altimeter (for landing procedure)
6. Artificial horizon
7. Variometer
8. Directional gyro
9. Board clock
10. Air pressure gauge
11. Output oil temperature gauge
12. Oil pressure gauge
13. Fuel pressure gauge
14. Boost pressure gauge
15. Tachometer
16. Auxiliary fuel tank valve lever
17. Three-way valve lever
18. Oil valve lever
19. Fire lever (Emergency lever)
20. Oil flow lever
21. Cooler lever

Master compass

Above: *Altimeter*
Right: *Integrated flight controller*
Below, right: *Variometer*
Below: *Artificial horizon*

Precision altimeter (for landing procedure)

Directional Gyro

Oil temperature gauge

Oil pressure gauge

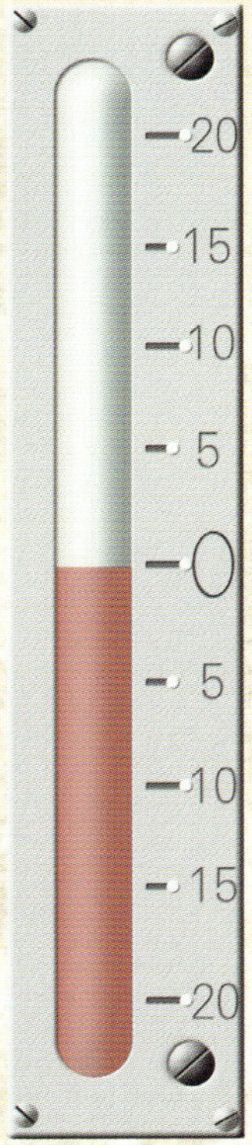

Fore-and-aft
level indicator

Fuel pressure gauge

Boost pressure gauge

Tachometer

Above: *PZL.23 Karaś rear cockpit.*

Left: *PZL.23 port side of the front cockpit.*

Right: *Starboard side of the observer's cockpit. Board clock and altimeter are visible in the top.*

(All photos from Technical Manual)

Lublin R-XIII

The Lublin R-XIII was the Polish army cooperation plane (observation and liaison plane), designed in the early-1930s in the Plage i Laśkiewicz factory in Lublin. It was the main army cooperation plane in the Invasion of Poland. Its variant Lublin R-XIV was a military trainer aircraft.

On 21 July 1931, 50 aircraft R-XIII were ordered. The first series of 30 were designated R-XIIIA, the next 20 were R-XIIIB. Both variants differed mostly in the type of a machine gun ring mounting. The first serial R-XIII was built on 7 June 1932. By 11 March 1933, all were given to the Air Force.

In 1932, the next 170 aircraft were ordered. 48 were built as the R-XIIIC variant with minor modifications, then 95 were built in the most numerous R-XIIID variant.

In 1931, one R-XIV was tested on floats, as a seaplane. Since tests came out well, the Polish Navy designated it R-XIII bis/hydro and ordered the next three (nos. 700-703). In 1933, the Navy ordered 10 R-XIII ter/hydro, which was a seaplane variant of R-XIIID (nos. 704-713). Finally, in 1934 the Navy bought 6 R-XIIIG seaplanes, differing in details from a previous variant – among others, a metal propeller (nos. 714-720). They were delivered by April 1935. All variants could also be easily converted to wheeled landing gear.

Right, above: Lublin R-XIII G pilot's instrument panel, photo from Technical instrument.

Right: Lublin R-XIII ter code no. 712. (A. Glass coll.)

Lublin R-XIII G

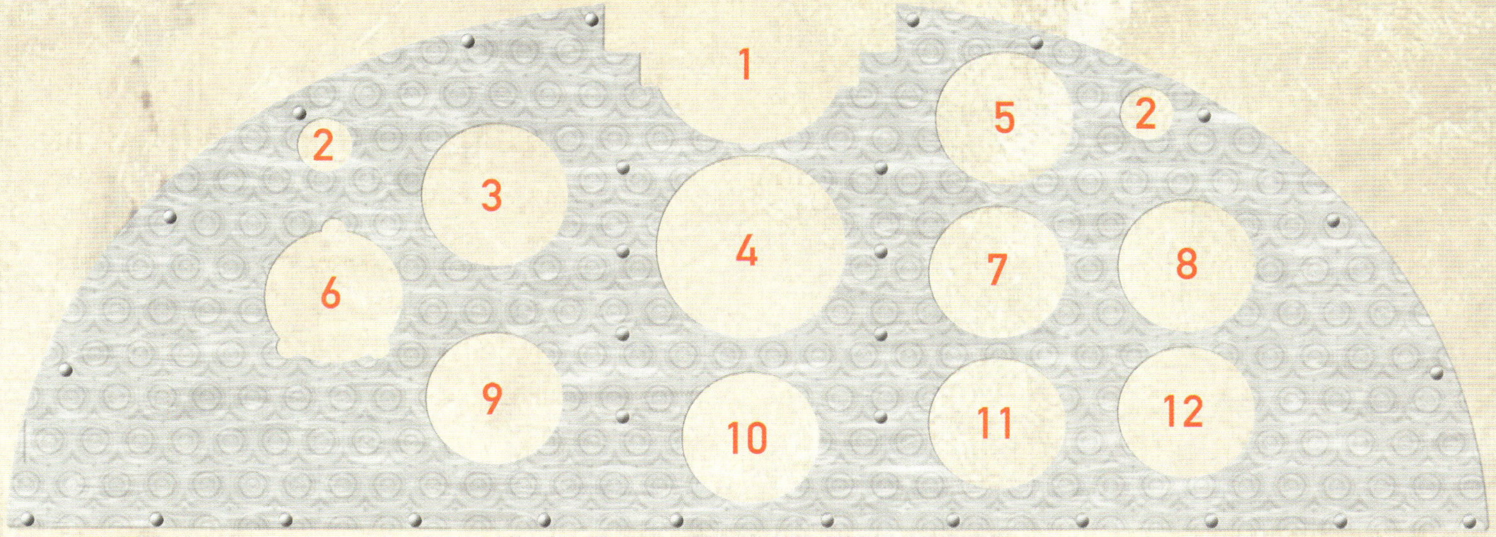

1. Master compass
2. Cockpit lamp's slots
3. Precision altimeter (for landing procedure)
4. Artificial horizon
5. Tachometer
6. Board clock
7. Oil temperature gauge
8. Oil pressure gauge
9. Altimeter
10. Airspeed indicator
11. Fuel contents gauge
12. Fuel pressure gauge

***Left:** Lublin R-XIII, port side of the fuselage with skin removed. Pilot's and gunner's compartments are visible. (Technical Manual)*

Master compass

Above: Precision altimeter (for landing procedure)

Precision altimeter (for landing procedure)

Below: Board clock

Left: Master compass

Right: Oil temperature gauge

Oil pressure gauge

Altimeter

Airspeed indicator

Left: Fuel contents gauge

Right: Fuel pressure gauge

Left: **Lublin R-XIII, starboard side of the fuselage with skin removed.** *Rear part of the fuselage, just behind the gunner's compartment.*

Left, bottom: *Gunner's compartment, rear part of the pilot's seat is visible.*

Right, bottom: *Port side of the pilot's compartment.*

(All photo Technical Manual)

RWD-14 Czapla

The RWD-14 Czapla (LWS Czapla), (in polish – heron) was a Polish army cooperation aircraft (observation, close reconnaissance and liaison aircraft), designed in the mid-1930s by the RWD team, and produced in the LWS factory from 1938. A series of 65 aircraft were built and most were used by the Polish Air Force observation squadrons during World War II in 1939.

The Czapla entered service in the Polish Air Force in the spring of 1939, equipping some observation squadrons (*eskadra obserwacyjna*). Due to its long development, it was not a modern aircraft, only a little better than the Lublin R-XIII. Its advantage was its short take-off (140 m) and landing (120 m), enabling it to operate from fields and meadows. Its modern successor, the LWS-3 Mewa (in polish – seagull), did not manage to enter operational units due to the war.

***Right:** RWD-14 pilot's cockpit, photo from Technical Manual.*

***Below:** RWD-14 during tests in 1938. (A. Glass coll.)*

RWD-14 Czapla

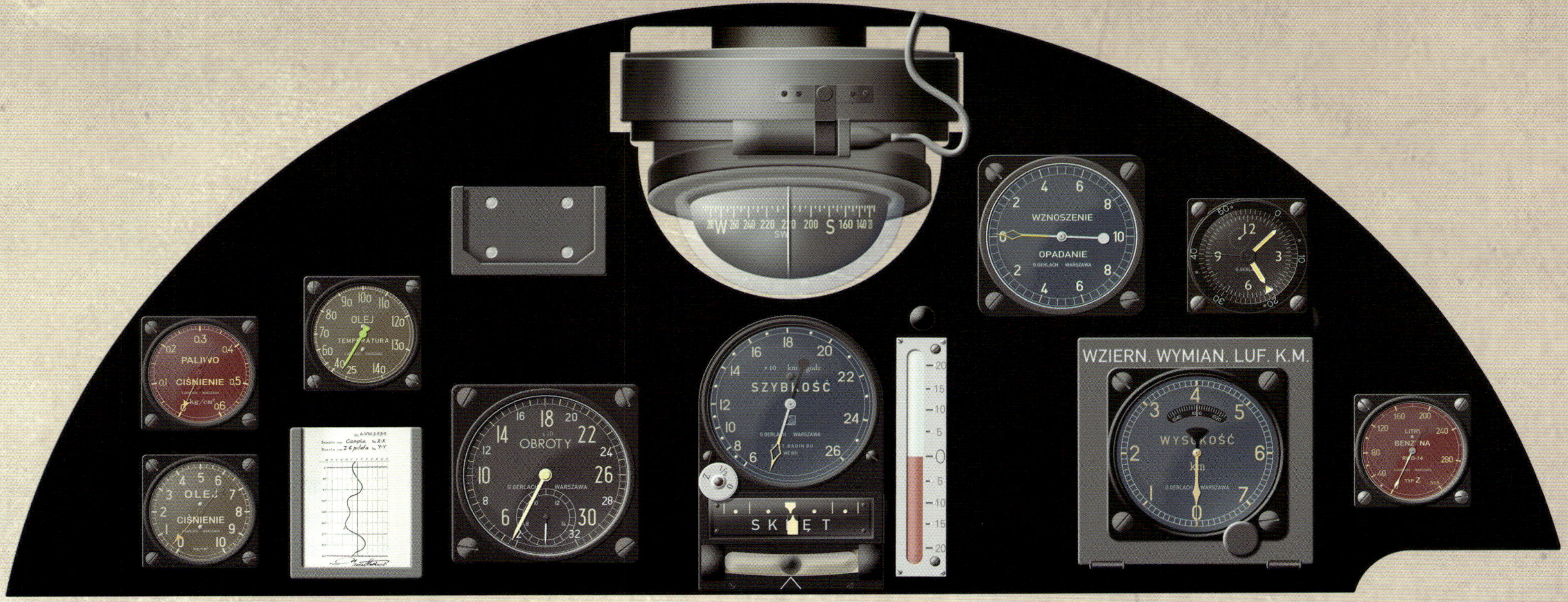

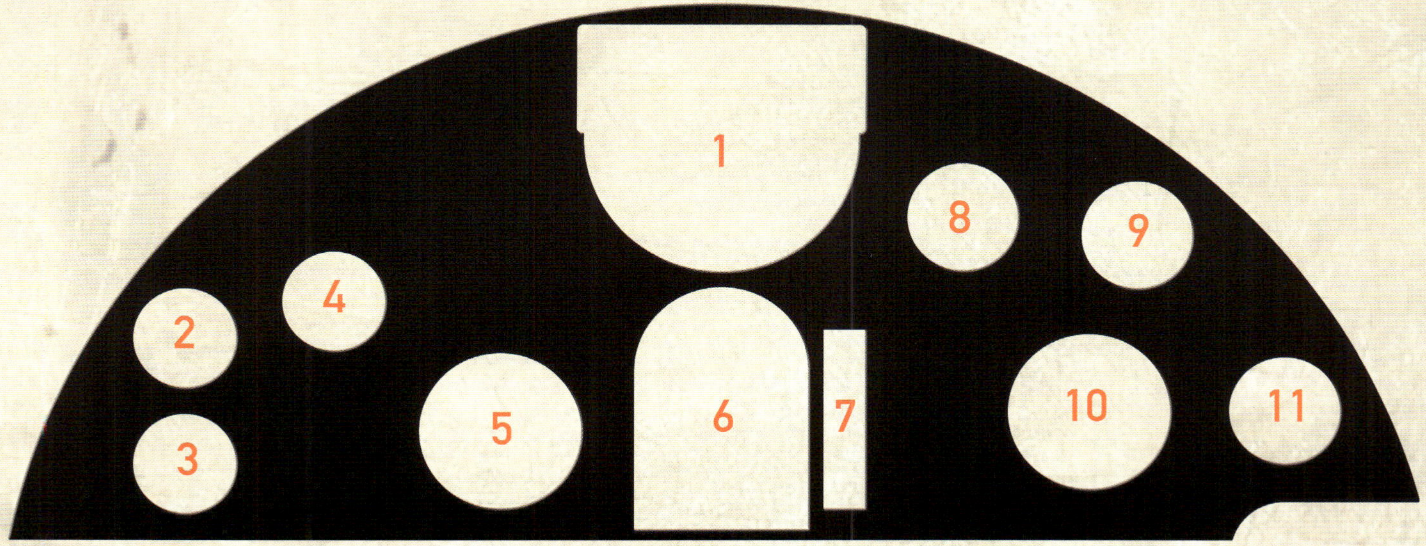

1. Master compass
2. Fuel pressure gauge
3. Oil pressure gauge
4. Oil temperature gauge
5. Tachometer
6. Integrated flight controller (airspeed indicator, turn and bank indicator)
7. Fore-and-aft level indicator
8. Variometer
9. Board clock.
10. Altimeter
11. Fuel contents gauge

Master compass

Fuel pressure gauge

Oil pressure gauge

Fuel contents gauge

Oil temperature gauge

Above: *Tachometer*
Right: *Integrated flight controller (Airspeed indicator, Turn and bank indicator)*

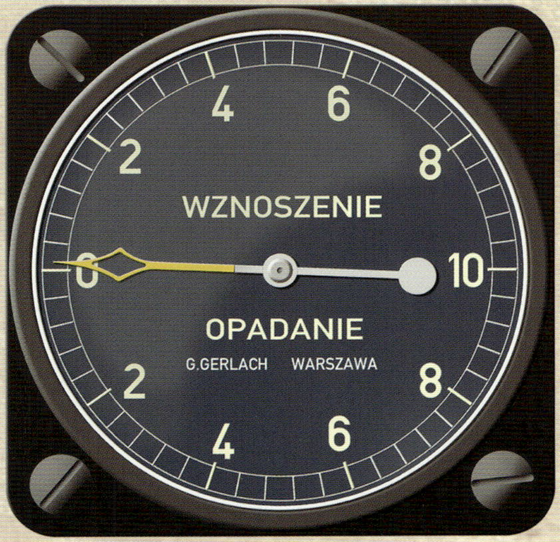

Above: *Variometer*
Left: *Fore-and-aft level indicator*

Above: *Altimeter*
Right: *Board clock*

PWS-26

The PWS-26 was a Polish advanced training aircraft, used from 1937 to 1939 by the Polish Air Force, constructed in the PWS (Podlaska Wytwórnia Samolotów – Podlasie Aircraft Factory). It was the second most numerous Polish pre-war aircraft, after the RWD-8.

Two-seater mixed construction (steel and wood) trainer biplane, canvas and plywood covered, with open cockpits. A fuselage of a metal frame, covered with wood and canvas, and partly with aluminium sheet. Two-spar rectangular wings with rounded tips, of wooden construction, canvas and plywood covered. Crew of two, sitting in tandem, with dual controls (instructor in front, student in rear). Some had an elevated canvas cover above the student's cockpit for blind flying training. Engine: 9-cylinder Wright Whirlwind J-5B air-cooled radial engine, built under license in the Polish Avia works. Two-blade wooden propeller of fixed pitch. Conventional landing gear, with a rear skid. Fuel tank 150 l in the fuselage. Armament (optional): 1 x 7.92 mm wz.33 or Vickers machine gun in the right-hand side of the fuselage, with interrupter gear, 1 x camera gun and 2 x 12 kg bombs.

Right: *Main, (rear) instrument panel of the PWS-26, photo from Technical Manual.*

Below: *PWS-26 prototype during tests. (A. Glass coll.)*

PWS-26 (rear cockpit)

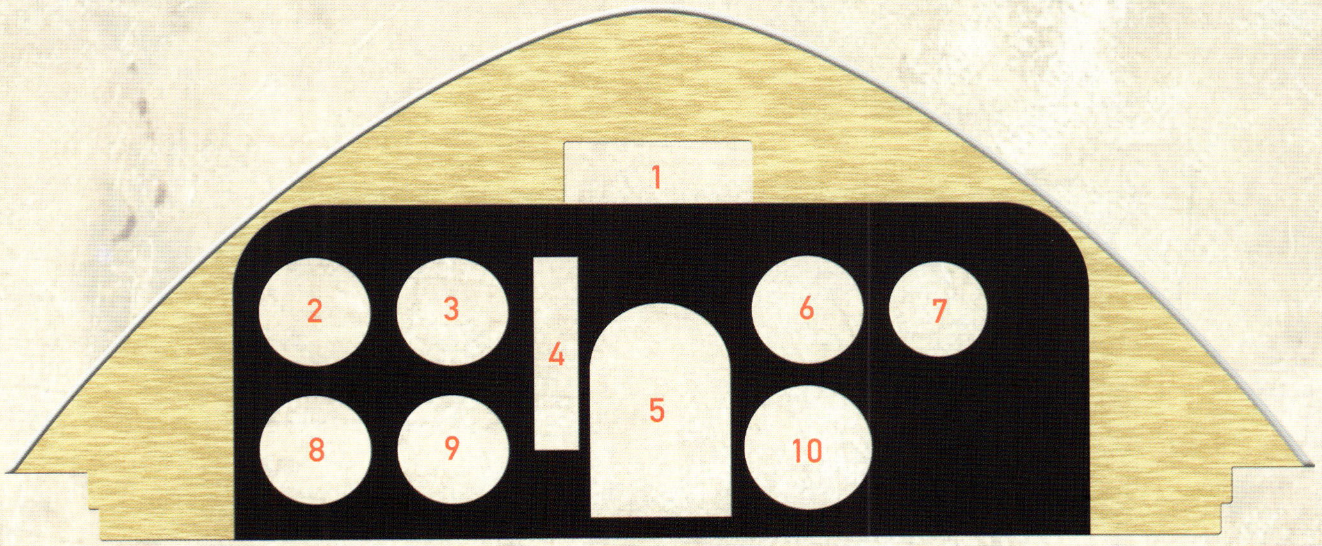

Rear cockpit (main)

1. Master compass
2. Oil temperature gauge
3. Oil pressure gauge
4. Fore-and-aft level indicator
5. Integrated flight controller (airspeed indicator, turn and bank indicator)
6. Board clock.
7. Air pressure gauge
8. Fuel contents gauge
9. Fuel pressure gauge
10. Altimeter

Front cockpit (instructor)

1. Fore-and-aft level indicator
2. Integrated flight controller (airspeed indicator, turn and bank indicator)
3. Board clock.
4. Altimeter
5. Tachometer (outside of fuselage, next to front cockpit)

PWS-26 (front)

Left: *Tachometer (outside of fuselage, next to front cockpit)*

Right: *Front (Instructor) instrument panel, photo from Technical Manual.*

Oil temperature gauge

Oil pressure gauge

Right: *Integrated flight controller (airspeed indicator, turn and bank indicator)*

Right, bottom: *Air pressure gauge*

Left: *Fore-and-aft level indicator*

Fuel pressure gauge

Fuel contents gauge

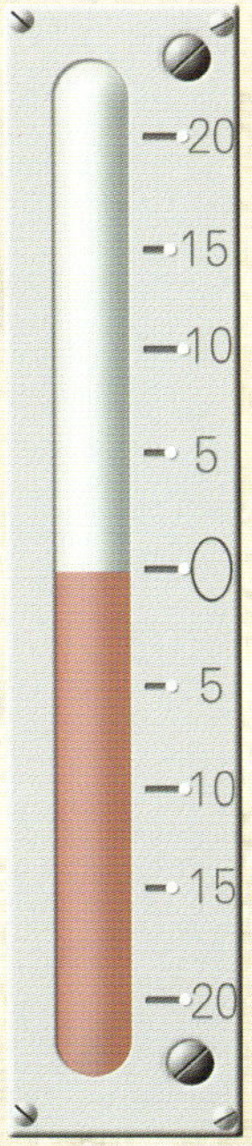

Left: Altimeter

Three photos of the PWS-26 preserved in Muzeum Lotnictwa Polskiego in Cracow. (Author)